201 Maths Activity Book

Can you add the given numbers?

16   9   15

Circle the greater number.

26   21   12

Solve 3x6 and circle the correct answer.

55   18   27

Wonder House

1. Circle the group on right with the same number of objects on left.

2. Trace numbers from 1-10.

3. Match the numbers with the correct objects.

4. What comes in the beginning?

5. Circle the group on right by adding 1 to the group on left.

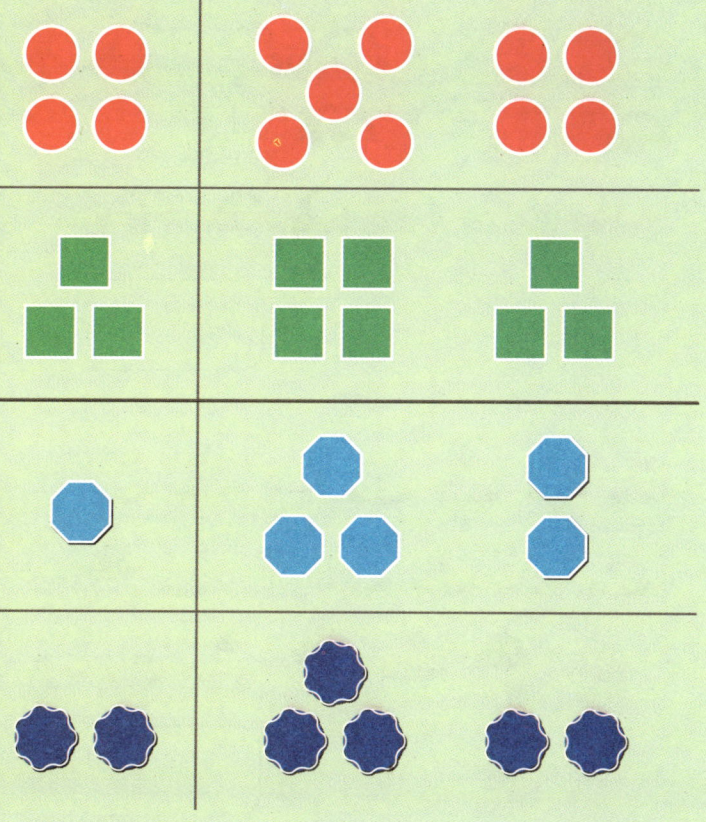

6. Use the color code at the bottom to color this basket of eggs.

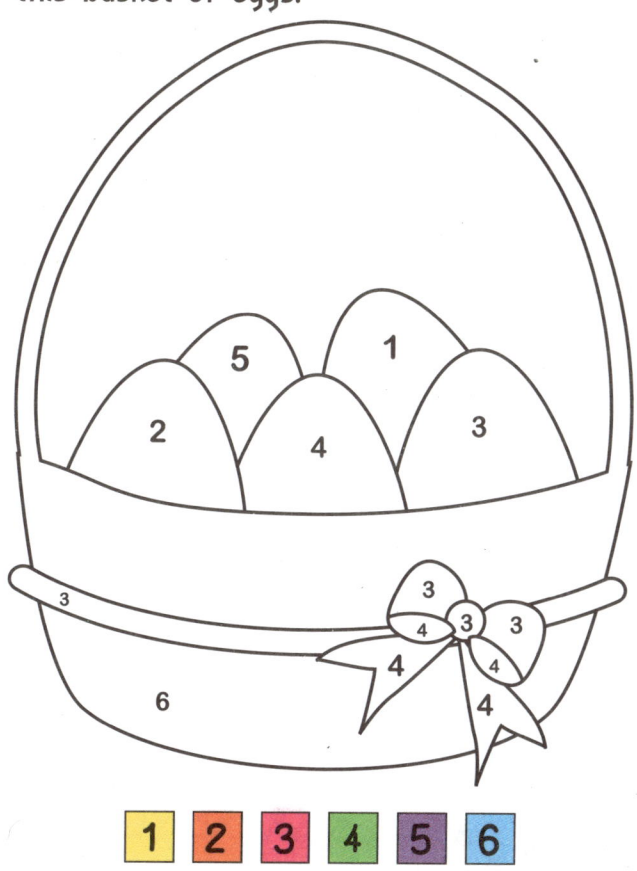

| 1 | 2 | 3 | 4 | 5 | 6 |

7. Count and write the number of cute ladybugs.

8. Count the objects given in the boxes and circle the correct answers.

6 7 8 9 5

2 8 3 6 10

9. Fill in the blank spaces to complete the number sequence. Trace the circles to draw a caterpillar.

10. How many similar objects are there? Write your answer in the box provided.

11. Count the butterflies and color the correct number.

6   11   9

12. Count the number of muffins and write the correct number in the box.

Answer: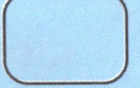

5. Circle the group on right by adding 1 to the group on left.

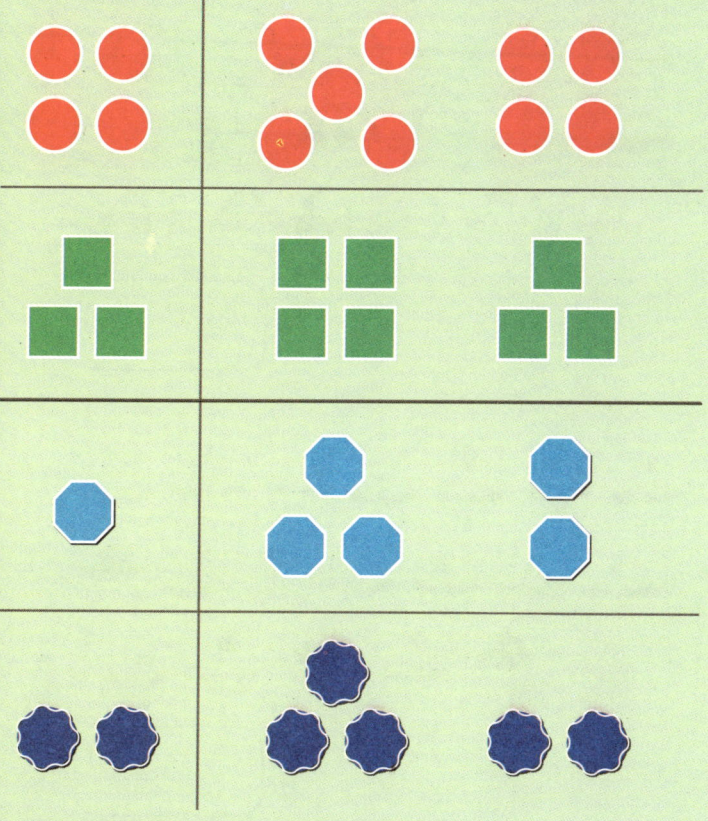

6. Use the color code at the bottom to color this basket of eggs.

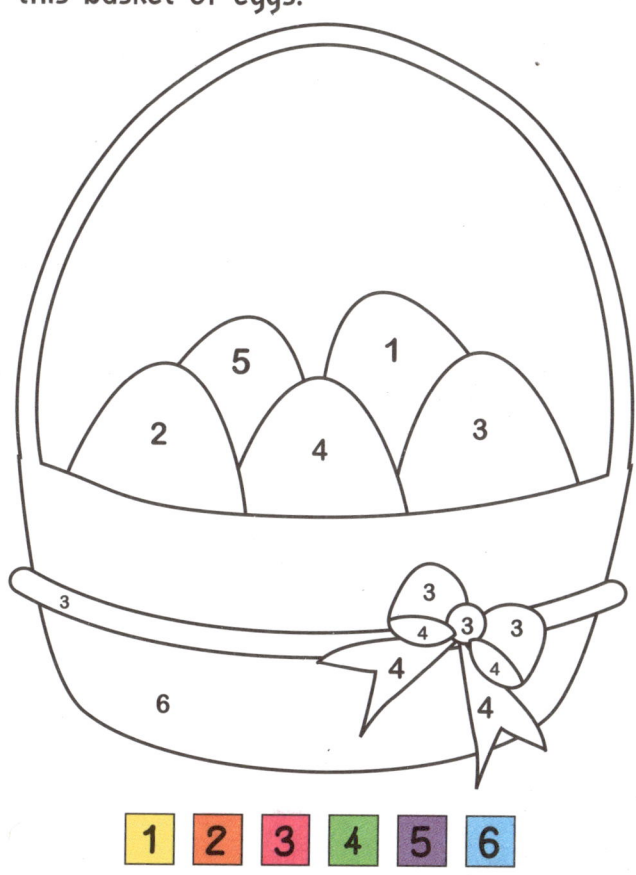

| 1 | 2 | 3 | 4 | 5 | 6 |

7. Count and write the number of cute ladybugs.

8. Count the objects given in the boxes and circle the correct answers.

6 7 8 9 5

2 8 3 6 10

9. Fill in the blank spaces to complete the number sequence. Trace the circles to draw a caterpillar.

10. How many similar objects are there? Write your answer in the box provided.

11. Count the butterflies and color the correct number.

6   11   9

12. Count the number of muffins and write the correct number in the box.

Answer: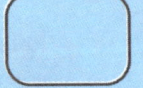

13. Count and write the number of animals in each box.

14. Count the gift boxes and circle the correct answer in the given cues.

15. What comes in between?

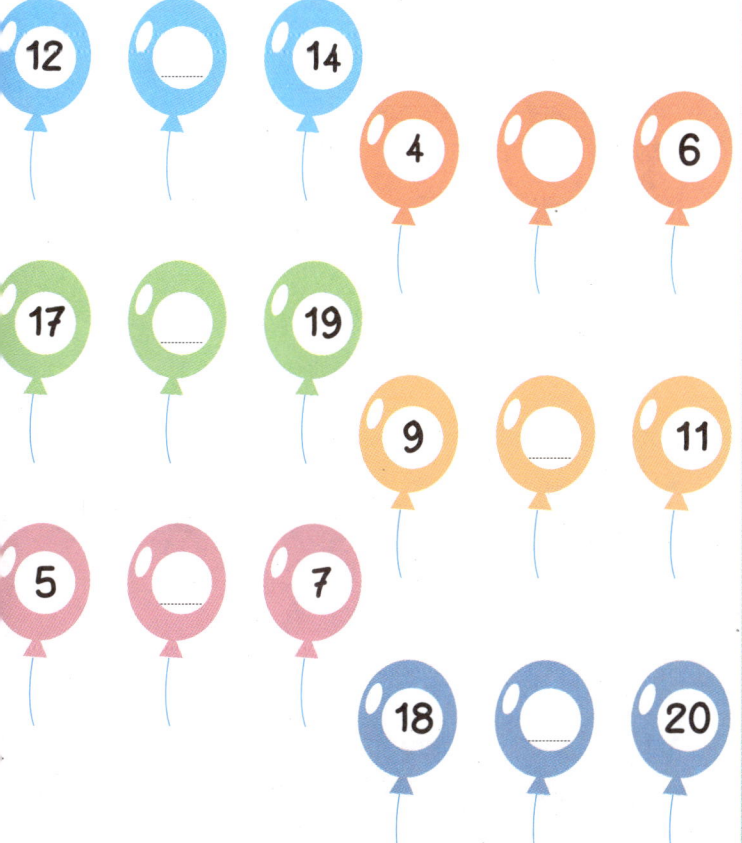

16. What comes after?

17. How many spots do the ladybugs have?

18. Place the jumbled numbers in ascending order.

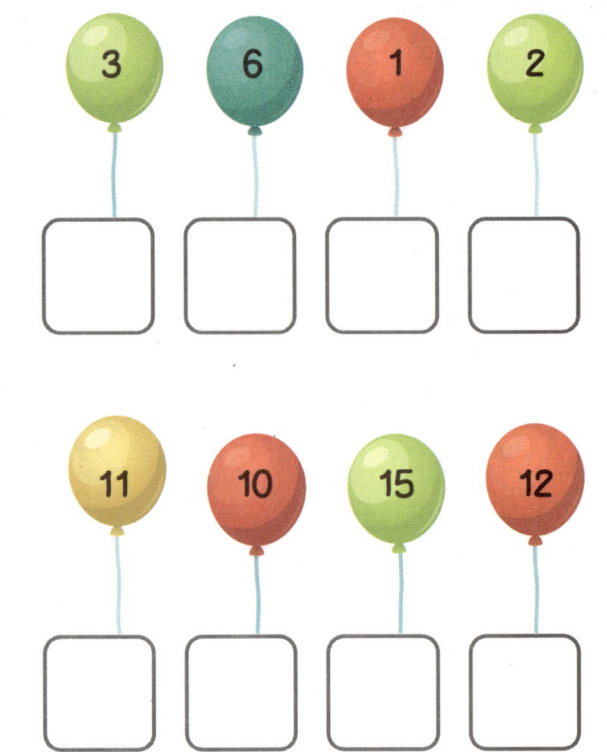

3   6   1   2

11   10   15   12

19. Place the jumbled numbers in descending order.

5   10   15   20

30   35   21   12

7   24   18   16

20. Write two numbers before and after the given numbers to complete the train.

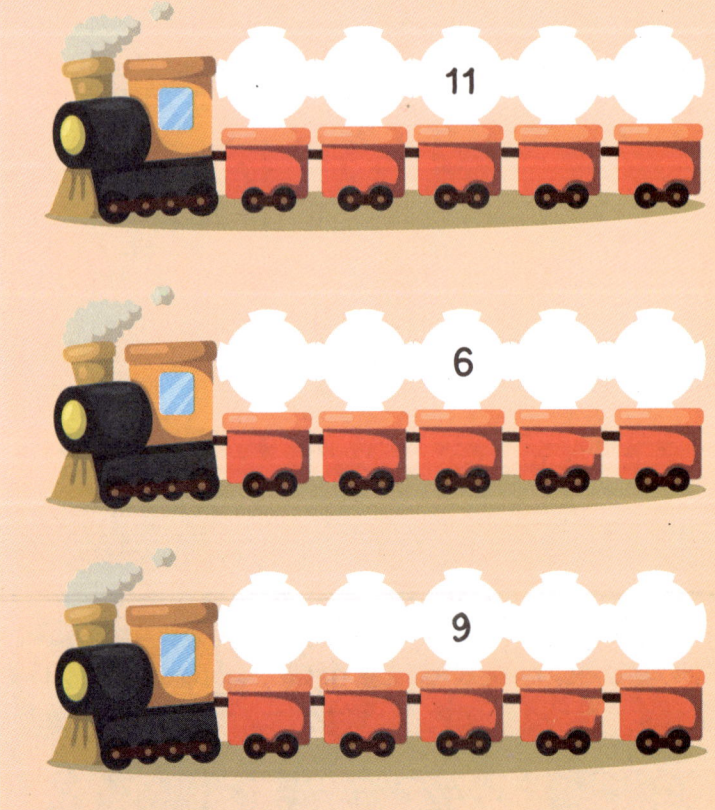

11

6

9

13. Count and write the number of animals in each box.

14. Count the gift boxes and circle the correct answer in the given cues.

15. What comes in between?

16. What comes after?

17. How many spots do the ladybugs have?

18. Place the jumbled numbers in ascending order.

Ascending order
Arranging numbers from smallest to largest.

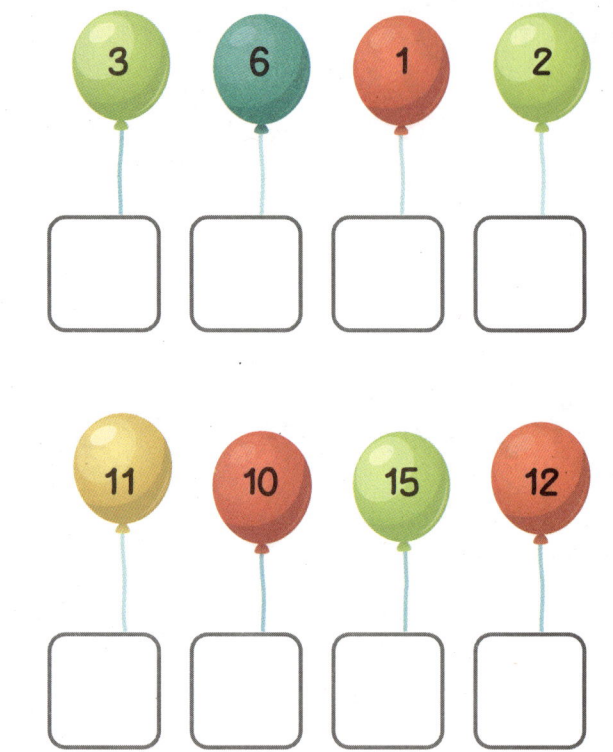

3   6   1   2

☐ ☐ ☐ ☐

11   10   15   12

☐ ☐ ☐ ☐

19. Place the jumbled numbers in descending order.

Descending order
Arranging numbers from largest to smallest.

5   10   15   20

⬡ ⬡ ⬡ ⬡

30   35   21   12

⬡ ⬡ ⬡ ⬡

7   24   18   16

⬡ ⬡ ⬡ ⬡

20. Write two numbers before and after the given numbers to complete the train.

11

6

9

6

21. Write the missing numbers.

Row 1: ☐, 2, ☐, 4, ☐, ☐
Row 2: ☐, ☐, 9, 10, ☐, ☐
Row 3: ☐, ☐, ☐, 16, ☐, ☐
Row 4: ☐, 20, ☐, ☐, 23, ☐

22. Match the number names with the correct numbers.

Twenty-two   **40**

Eleven   **59**

Forty   **22**

Fifty-nine   **6**

Eighty   **11**

Six   **80**

23. Round off the following numbers.

89   43   22

☐   ☐   ☐

159   201   333

☐   ☐   ☐

24. Use the correct sign (<, >, =).

23 • • 25

11 • • 12

10 • • 10

8   • • 9

49 • • 56

7

25. Write the numbers in ascending order.

Ordinal numbers tell the position or rank of something in a list.
For e.g., 1st (first), 2nd (second), etc.

26. Follow the instructions and color the chicks.

Color the **first** chick yellow.

Color the **second** chick orange.

Color the **third** chick red.

Color the **fourth** chick pink.

Color the **fifth** chick purple.

Even numbers are divisible by 2 without remainders. They end in 0, 2, 4, 6, or 8. Odd numbers are not evenly divisible by 2 and end in 1, 3, 5, 7, or 9.

27. Color the flowers with even numbers.

| 2 | 6 | 7 | 10 | 11 | 20 |

28. Color the bugs with odd numbers.

Place value is the value of each digit in a number.
For e.g., the 5 in 50 has a place value of 5 tens or 50.

29. I am an odd number.
I am between 30 & 40.
My ones digit is 7.

I am

------- | -------

30. Circle the greater number in each pair.

3   7   9   6   15   8

11   5   4   9   20   15

3   10   45   54   6   2

31. Choose the correct option.

a. 1   6   4   Count my legs.

b. 9   5   2   Count the flowers.

c. 7   3   5   Count my spots.

32. Color each shape, count its sides and tick the correct number.

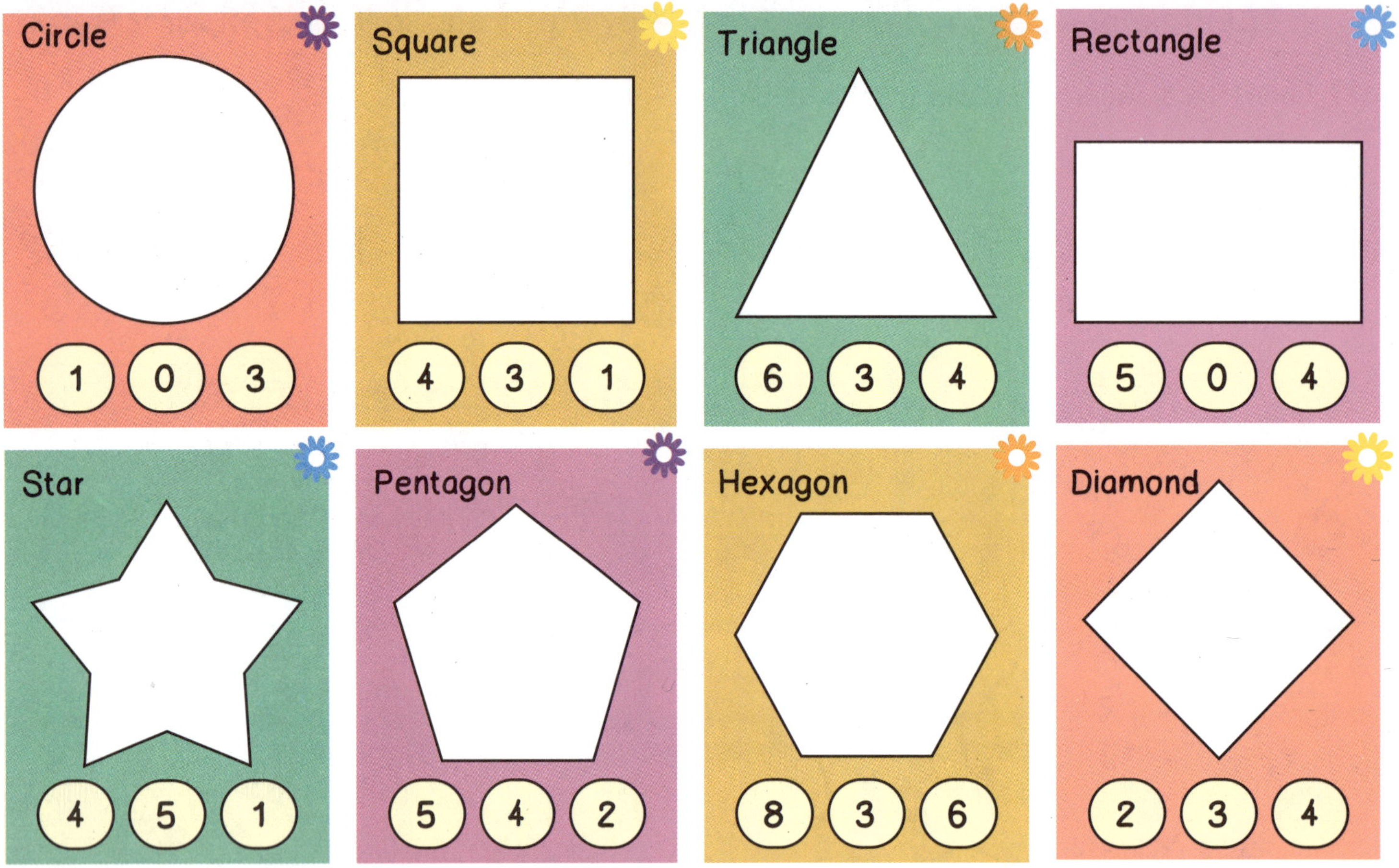

Circle

(1) (0) (3)

Square

(4) (3) (1)

Triangle

(6) (3) (4)

Rectangle

(5) (0) (4)

Star

(4) (5) (1)

Pentagon

(5) (4) (2)

Hexagon

(8) (3) (6)

Diamond

(2) (3) (4)

33. Which shape are the following objects?

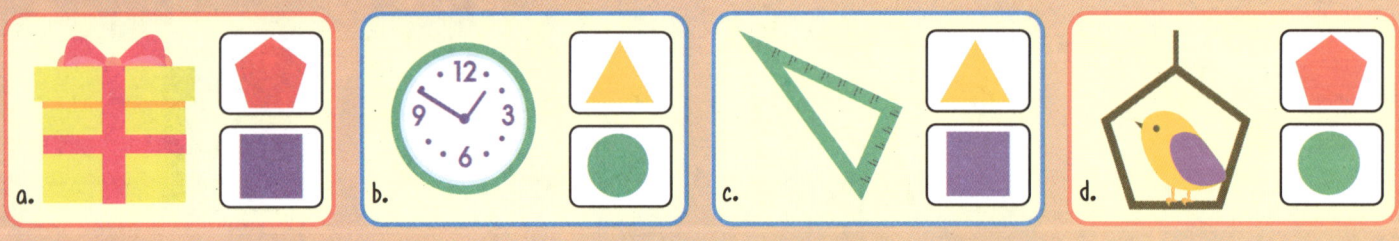

a.   b.   c.   d.

34. Color the butterfly as per the color code.

35. Complete the pattern.

36. Color the shapes and fill the graph.

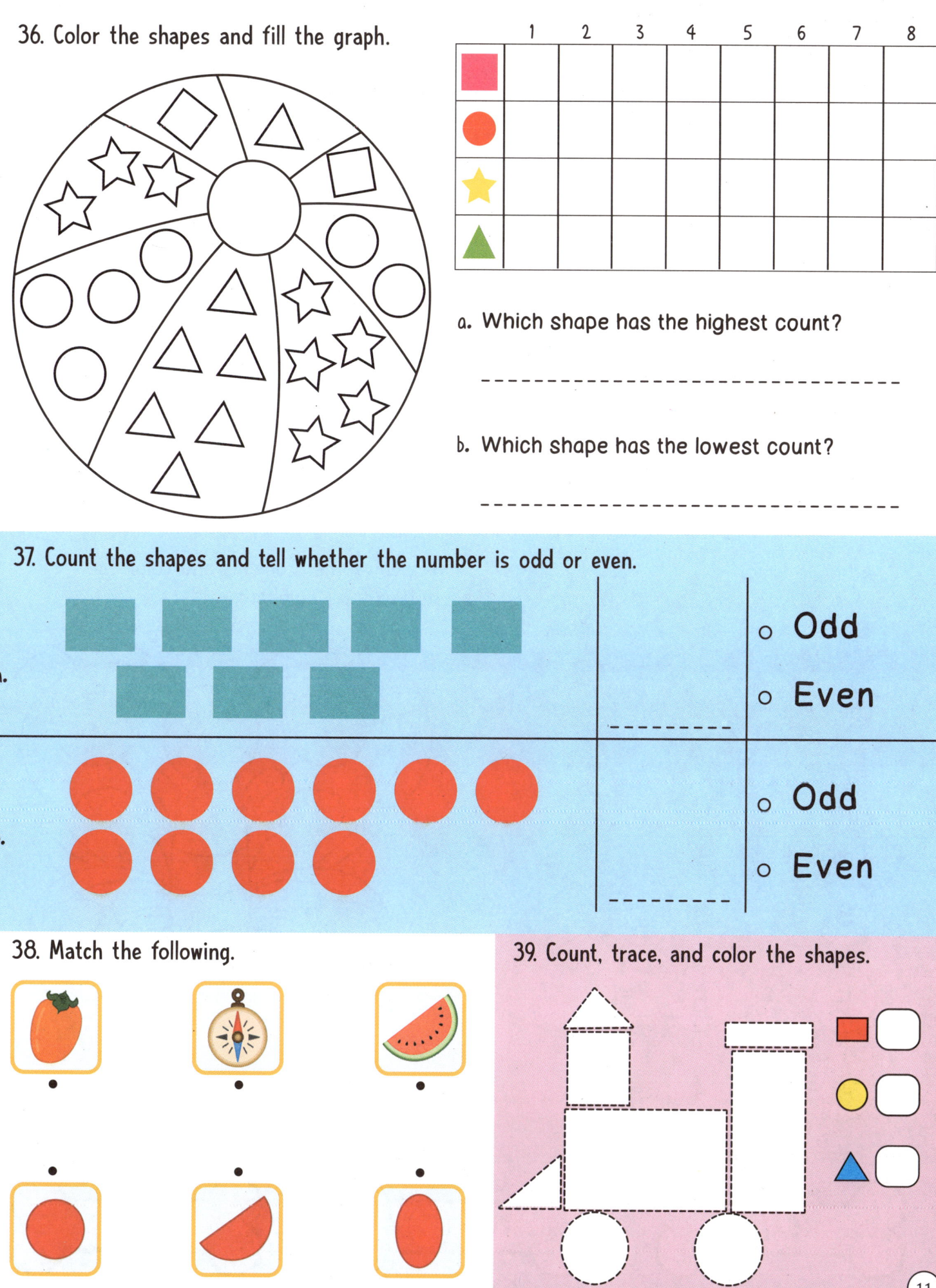

	1	2	3	4	5	6	7	8
🟥								
🔴								
⭐								
🔺								

a. Which shape has the highest count?

_____

b. Which shape has the lowest count?

_____

37. Count the shapes and tell whether the number is odd or even.

a.

o Odd

o Even

_____

b.

o Odd

o Even

_____

38. Match the following.

39. Count, trace, and color the shapes.

11

40. Add the leaves and write the answers.

41. Add the fruits and vegetables. Write the answers in the box.

42. Add the twos.

2 + 2 + 2 + 2 + 2 + 2 =

43. Solve the problems and color the butterfly by numbers.

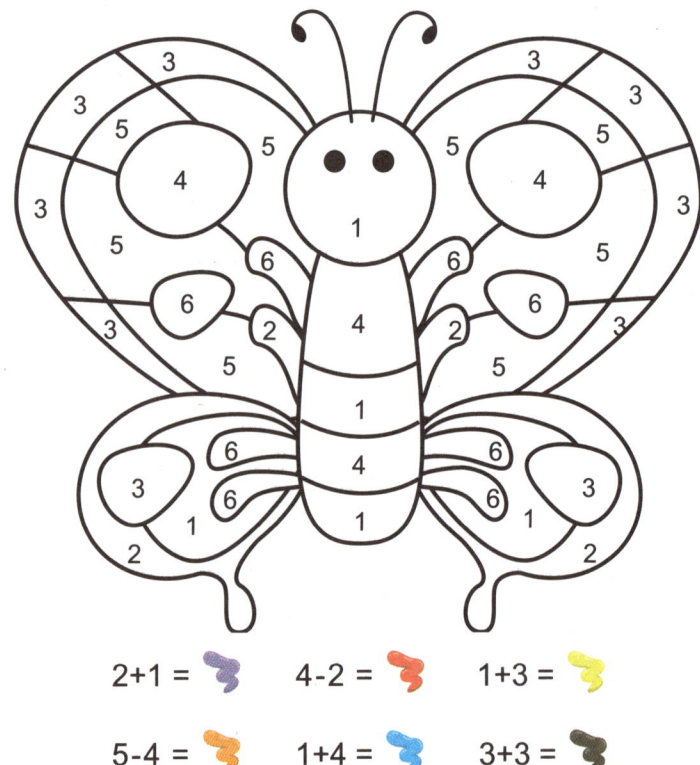

2+1 =   4-2 =   1+3 =

5-4 =   1+4 =   3+3 =

44. Double it up!

3	+	3	=	
5	+	5	=	
4	+	4	=	

45. Color the bone with even answer.

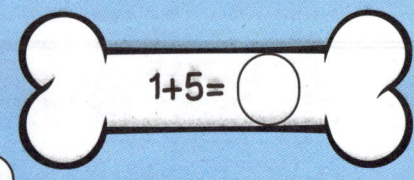
1+5=

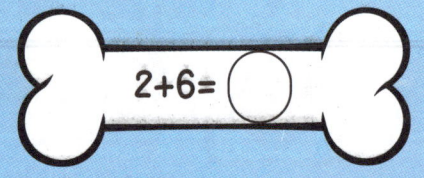
2+6=

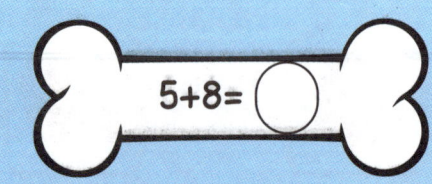
5+8=

12

46. Solve the following sums.

 $+$ $=$ ☐

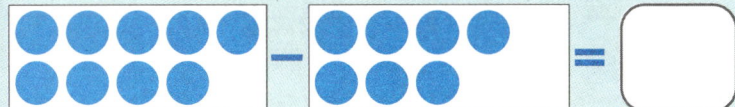 $-$ $=$ ☐

 $+$ $=$ ☐

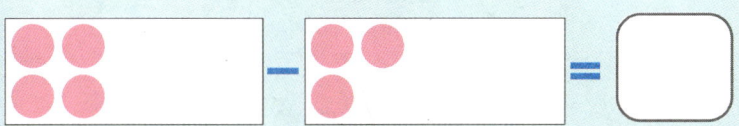 $-$ $=$ ☐

47. Color the boxes blue if the sum is 12.

1	0	0	8	1	1
+	2	+	2	+	1

☐   ☐   ☐

1	4	0	6	0	7
+	2	+	6	+	5

☐   ☐   ☐

48. Add the ice cream cones and write their sum in the scoop.

5 $+$ 2 $+$ 6 $=$

1 $+$ 4 $+$ 2 $=$

49. Fill in the blanks with the correct numbers.

10 $+$ 2 $=$

5 $+$ 6 $=$

16 $+$ 1 $=$

6 $+$ 8 $=$

50. Put a (✓) in front of correct sums.

16 $+$ 9 $=$ 30 ☐

10 $+$ 4 $=$ 14 ☐

9 $+$ 3 $=$ 12 ☐

12 $+$ 2 $=$ 15 ☐

18 $+$ 2 $=$ 20 ☐

20 $+$ 5 $=$ 27 ☐

51. Subtract the given images and write the answer.

52. Count all the dinosaurs that are not crossed and circle the correct number.

53. Subtract using the number line. Draw the jumps that you make and write the answers.

5-2=

8-7=

54. Fill the missing numbers by subtracting 3 from each tomato.

55. Find the difference and color the number of eggs in the answer.

18 - 12 =

46. Solve the following sums.

 $+$ $=$ ☐

 $-$ $=$ ☐

 $+$ $=$ ☐

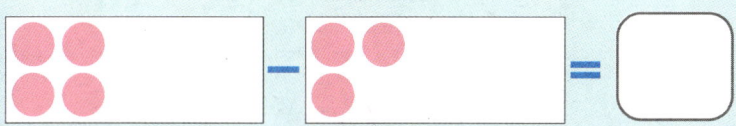 $-$ $=$ ☐

47. Color the boxes blue if the sum is 12.

1 0	0 8	1 1
+ 2	+ 2	+ 1
☐	☐	☐

1 4	0 6	0 7
+ 2	+ 6	+ 5
☐	☐	☐

48. Add the ice cream cones and write their sum in the scoop.

5 + 2 + 6 =    1 + 4 + 2 =

49. Fill in the blanks with the correct numbers.

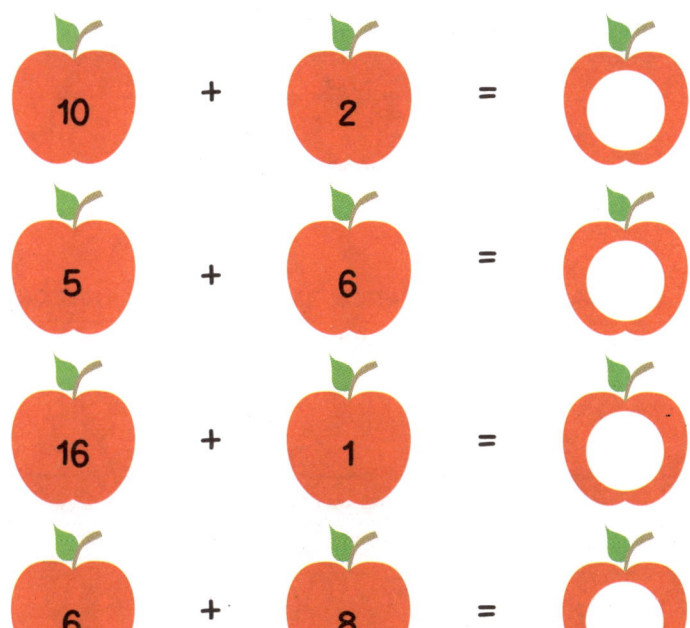

10 + 2 =

5 + 6 =

16 + 1 =

6 + 8 =

50. Put a (✓) in front of correct sums.

16 + 9 = 30   ☐

10 + 4 = 14

9 + 3 = 12

12 + 2 = 15

18 + 2 = 20

20 + 5 = 27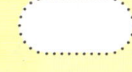

13

51. Subtract the given images and write the answer.

52. Count all the dinosaurs that are not crossed and circle the correct number.

4   6   3   8

1   3   5   2

53. Subtract using the number line. Draw the jumps that you make and write the answers.

5-2=

8-7=

54. Fill the missing numbers by subtracting 3 from each tomato.

55. Find the difference and color the number of eggs in the answer.

18 - 12 =

56. Subtract the given numbers on the watermelon slices and write the answers.

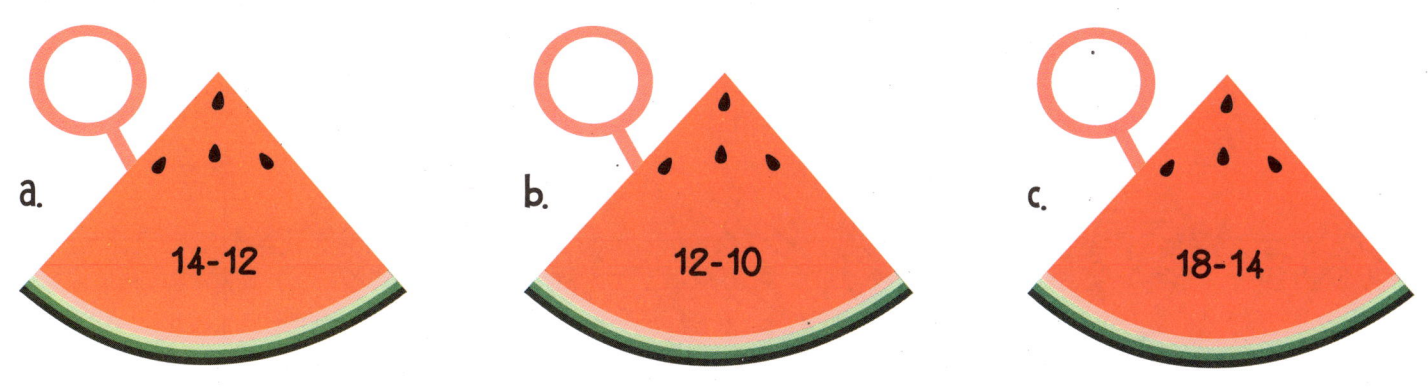

a. 14-12   b. 12-10   c. 18-14

57. Fill in the missing numbers by subtracting 2 each time.

a. 20   ___   16   ___   12

b. 15   ___   11   ___   7

58. Subtract correctly to make the UFOs go away.

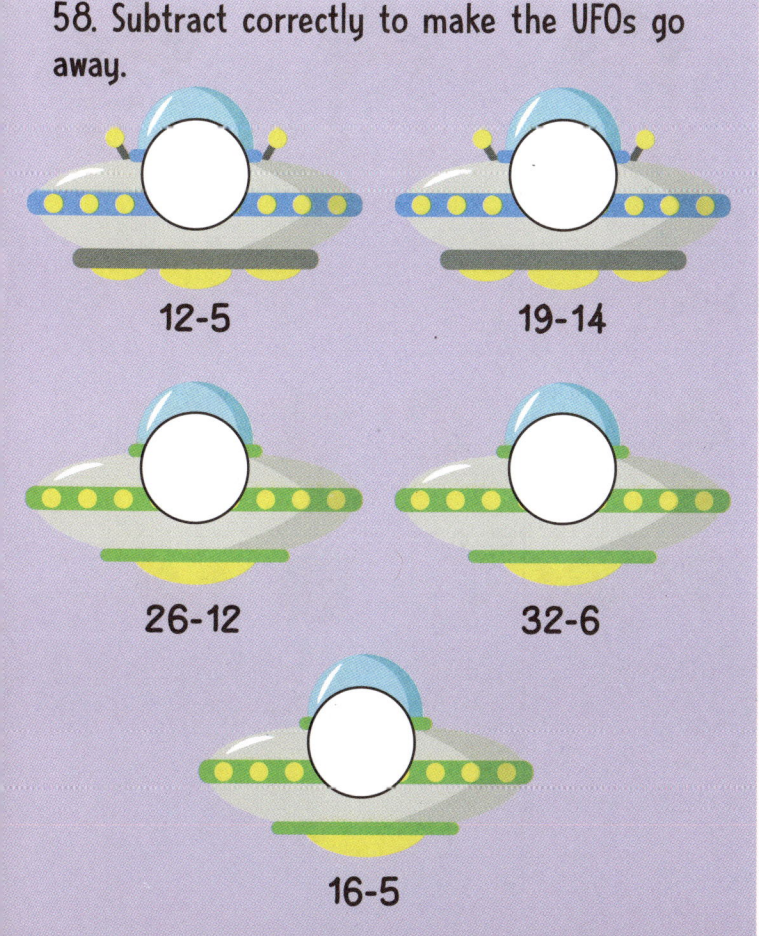

12-5   19-14

26-12   32-6

16-5

59. Use the color code to color the engine.

● 3-2   ● 6-4   ● 7-5   ● 8-4

(15)

60. Help the bunny move forward by drawing the jumps on the number line.

1 2 3 4 5 6 7 8 9

1+3=

[ ]

1 2 3 4 5 6 7 8 9

5+3=

[ ]

61. Add 3 to each bell as you jump and complete the row.

3   ◯   9   ◯   ◯   18

62. Add the given images and circle the correct answer.

+   +   +

3+3=   4+3=   3+2=

8 6 5   7 5 2   1 8 5

16

63. Count and subtract. Write the answers in the given boxes.

a.

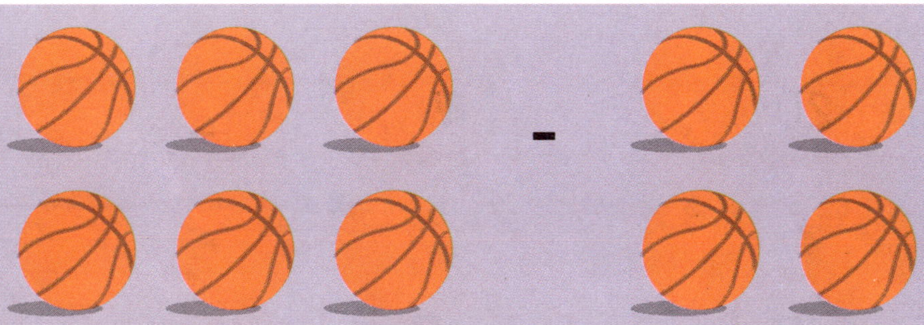

☐ - ☐

☐

b.

☐ - ☐

☐

64. Subtract the given numbers and write the difference in the space provided.

a. $13-5=$

b. $9-7=$

c. $18-4=$

d. $16-3=$

65. Subtract the numbers and write the difference in the empty circles.

a. $12-5$

b. $19-9$

c. $12-3$

66. Fill in the blanks with the correct numbers.

10 + _ = 20

7 + _ = 14

5 + _ = 12

15 + _ = 17

67. Add all the numbers given on the balls and write the answer.

68. Fill in the blanks with numbers that add up to 40.

..... + 15 = 40

..... + 19 = 40

14 + ... = 40

32 + ... = 40

69. Subtract the following and write the answer in the box.

18

70. Add the objects and write the sum.

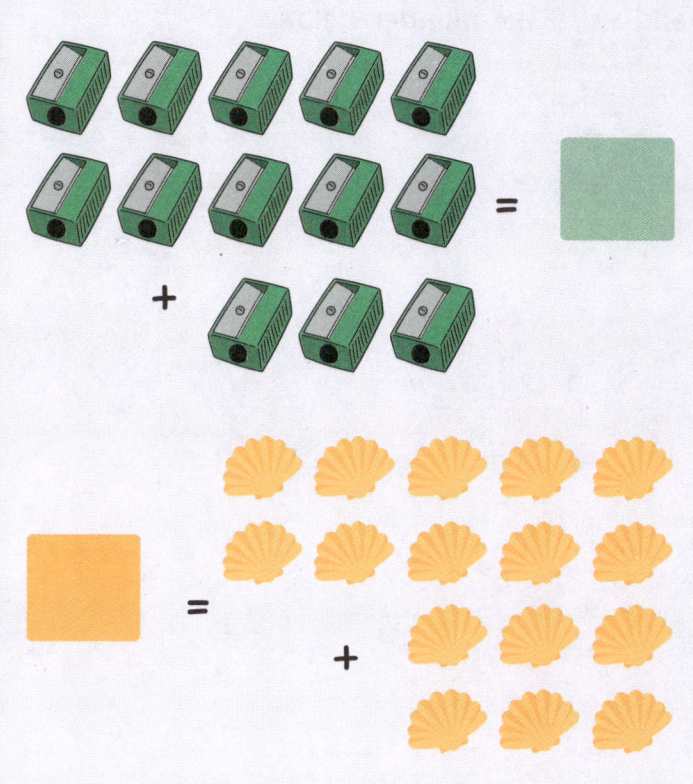

71. Subtract the following.

35 - 17 = ...

50 - 20 = ...

19 - 10 = ...

65 - 27 = ...

15 - 12 = ...

72. Solve the problem.

Tanya had 57 pears.

She gave 31 pears to her friend.

How many pears is Tanya left with now?

Answer:

73. Add the numbers.

2 + 2 + 2 + 2

= 2 + 2

3 + 3 + 3 + 3

= 3 + 3

74. Fill in the numbers that come before and after the given number.

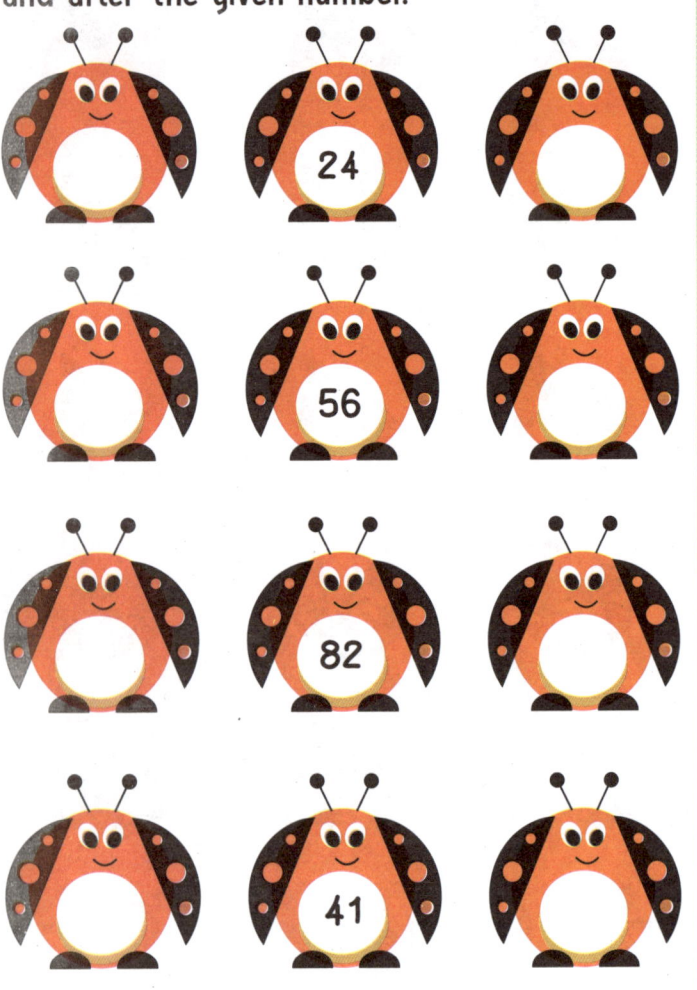

	24	
	56	
	82	
	41	

75. 5 seeds were planted in each pot. How many still need to grow? Draw the flowers and write the number below.

76. Draw each bee's route to the purple flower by following the instructions.

a. Add 2 every time

4   1   6   10   13   16
5   14   17
10   12   7
8

b. Add 3 every time

9
6   14   15   21   31
8
13   6   12   11   18   24

20

77. There are 6 bees in the beehive. 3 more bees join them. How many bees are there now?

78. There were 5 pears on the tree. 1 more pear grew on the tree. How many pears are there on the tree now?

79. There were 8 eggs in the crate. The farmer bought 2 more eggs. How many eggs are there in the crate now?

80. Fill in the missing numbers.

22 ___ 24 ___ 26

51 ___ ___ ___ 55

81. Skip count by 5 and write the numbers that follow.

5 ___ ___ ___

82. Arrange the following numbers in descending order.

71 73 75 70

122 125 120 122

83. Study the abacus and write the answers.

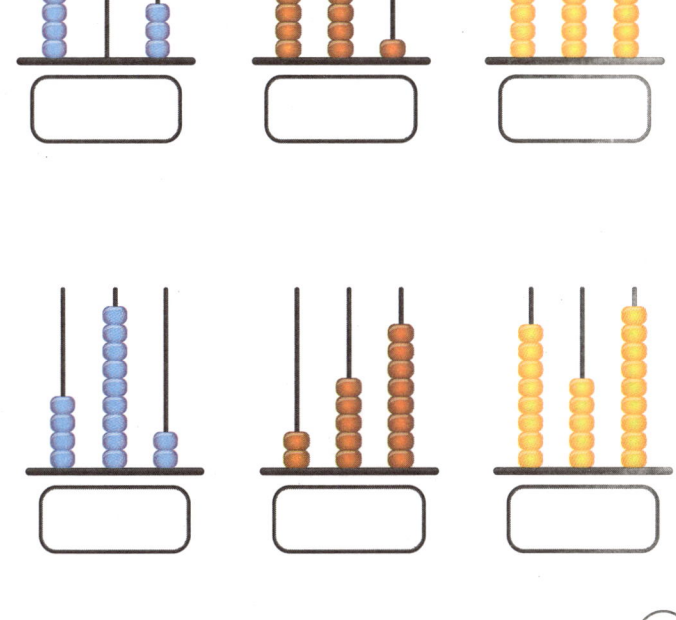

84. Write the expanded form of the numbers along with their number names and place values.

986 I am expanding.

My number name is

_____

I have

..... hundreds ..... tens .....ones

_____

My expanded form is

-------- + -------- + --------

116 I am expanding.

My number name is

_____

I have

.....hundreds ..... tens .....ones

_____

My expanded form is

-------- + -------- + --------

208 I am expanding.

My number name is

_____

I have

.....hundreds ..... tens .....ones

_____

My expanded form is

-------- + -------- + --------

85. Write the place value of the underlined digits.

3̲49   ----------   61̲4   ----------   4̲90   ----------   25̲4   ----------

86. Color the clouds with even numbers.

16   5   1   3

6   10   1   3

9   2   17   12

14   13   4   19

87. Color the flowers with odd numbers.

11   21

8   25   12

13   16

88. If "O" is 5, "W" is 2, and "L" is 6, what is the total of-

O   W   L

+   +

TOTAL

---

89. Fill in the boxes with the missing numbers between 1 & 50.

1	2		4	5
6				10
11				15
		18	19	20
	22	23		

	27	28	29	30
31	32			
	37	38	39	
	42	43	44	
		48		

---

90. Skip count by 5 and complete the sequence.

5   10   15   -----

-----   35   40   -----

-----   60   -----   70

-----   -----   90   95

---

91. Skip count by 2 in reverse order and reach the end.

100 — 98 — 96 — 94

76

23

92. Color the cupcake with the greatest number.

93. Fill in the missing numbers on the ice cream scoops. The bottom scoops add up to the number on the top scoop.

a.  $2 + 5$

b.  $8 + 4$

c.  9 $+ 7$

d.  6 $1 +$

e.  $2 + 3$

94. Solve the following.

 = 3   = 2    = 4   = 6

a.  −  = ☐

b.  −  = ☐

c.  −  = ☐

d.  −  = ☐

95. Add the numbers and write the answer in the given space.

a.

b.

96. Count and write the number of odds and evens in the jar.

Odd ◯

Even ◯

24

97. Solve and write the missing numbers.

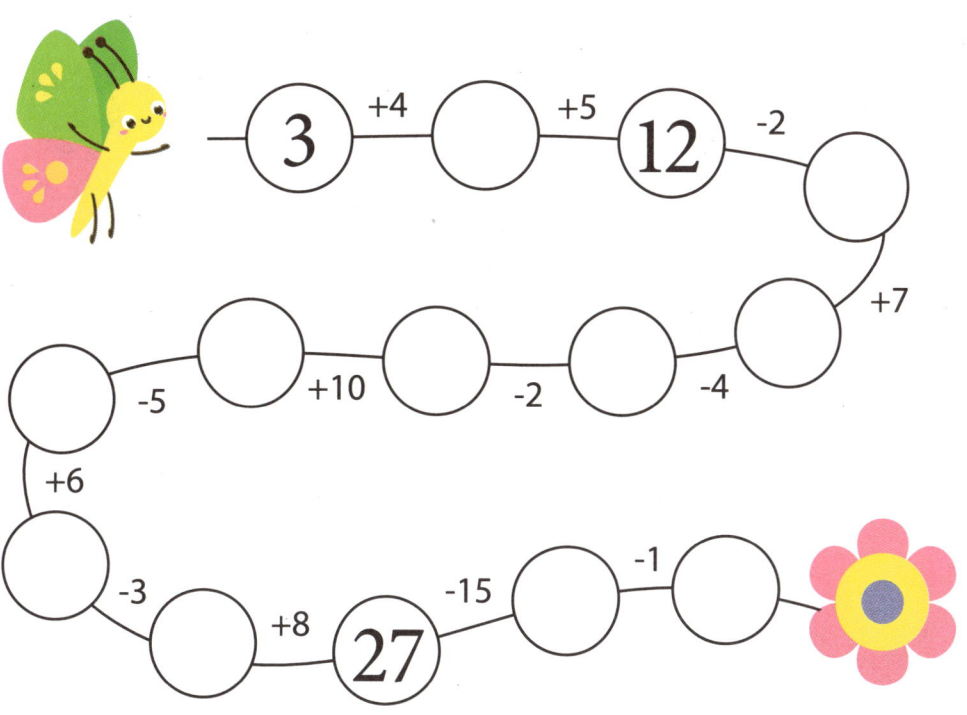

98. If it's true, color 🟩.
If it's false, color 🟧.

a. | 3 | = | 3 | ☐
b. | 5 | < | 4 | ☐
c. | 6 | < | 8 | ☐
d. | 1 | > | 4 | ☐
e. | 9 | = | 7 | ☐

99. Write the number that comes out of the robot.

100. Balance the scales.

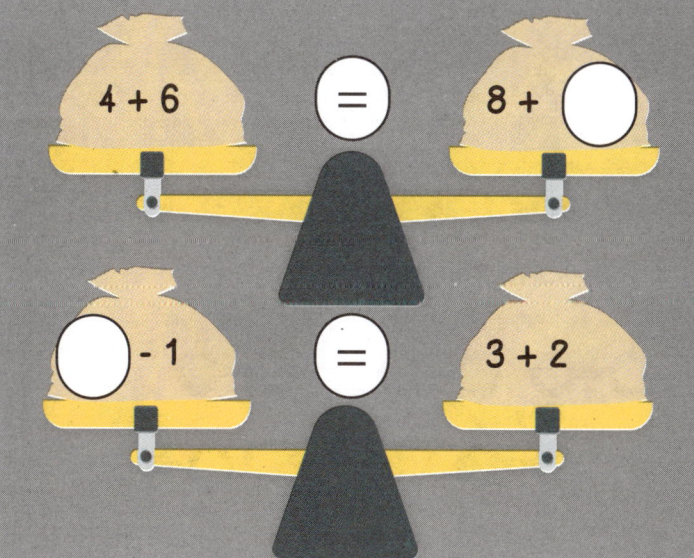

101. Solve the problem train from left to right.

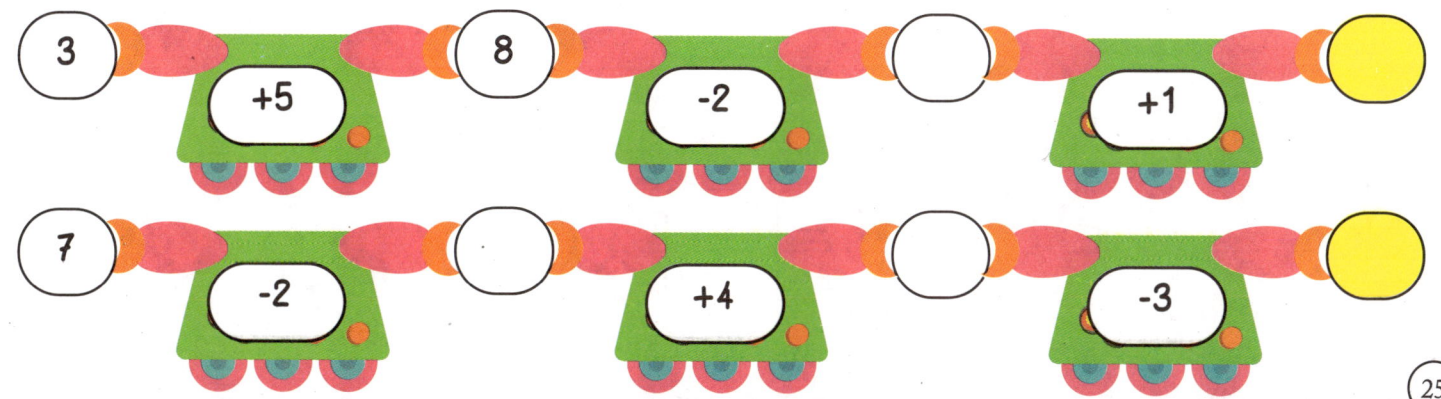

25

102. Multiply the row and column to complete the tables.

×	1	2	3	4	5	6	7	8	9	10
2			6							
3						18				
4										40
5				20						
6							42			
7									63	
8		16								
9					45					
10								80		

103. Solve the sums.

a. [trucks] X [trucks] = $\begin{array}{c} 3 \times 2 \\ = 6 \end{array}$

b. [submarines] X [submarines] = [ ]

c. [rockets] X [rockets] = [ ]

104. Solve the multiplication problems.

5 X 7 =

9 X 3 =

4 X 8 =

26

105. Write the missing numbers.

2 × ⬜ = 10   10 × 2 = ⬜

106. Match the multiplications with their answers.

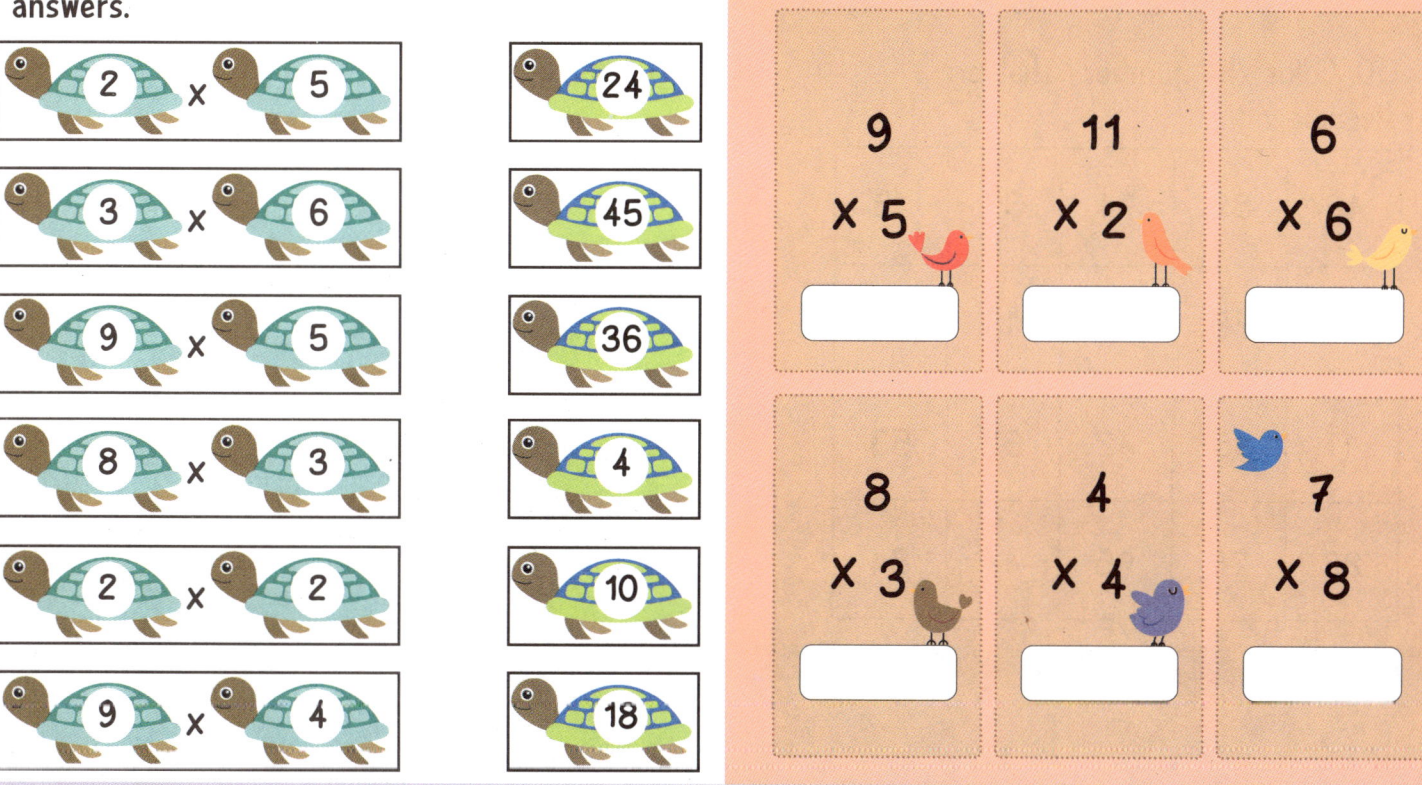

2 × 5   24

3 × 6   45

9 × 5   36

8 × 3   4

2 × 2   10

9 × 4   18

107. Solve the multiplication problems.

9 × 5	11 × 2	6 × 6
⬜	⬜	⬜
8 × 3	4 × 4	7 × 8
⬜	⬜	⬜

108. Multiply the numbers and write answers on each coach.

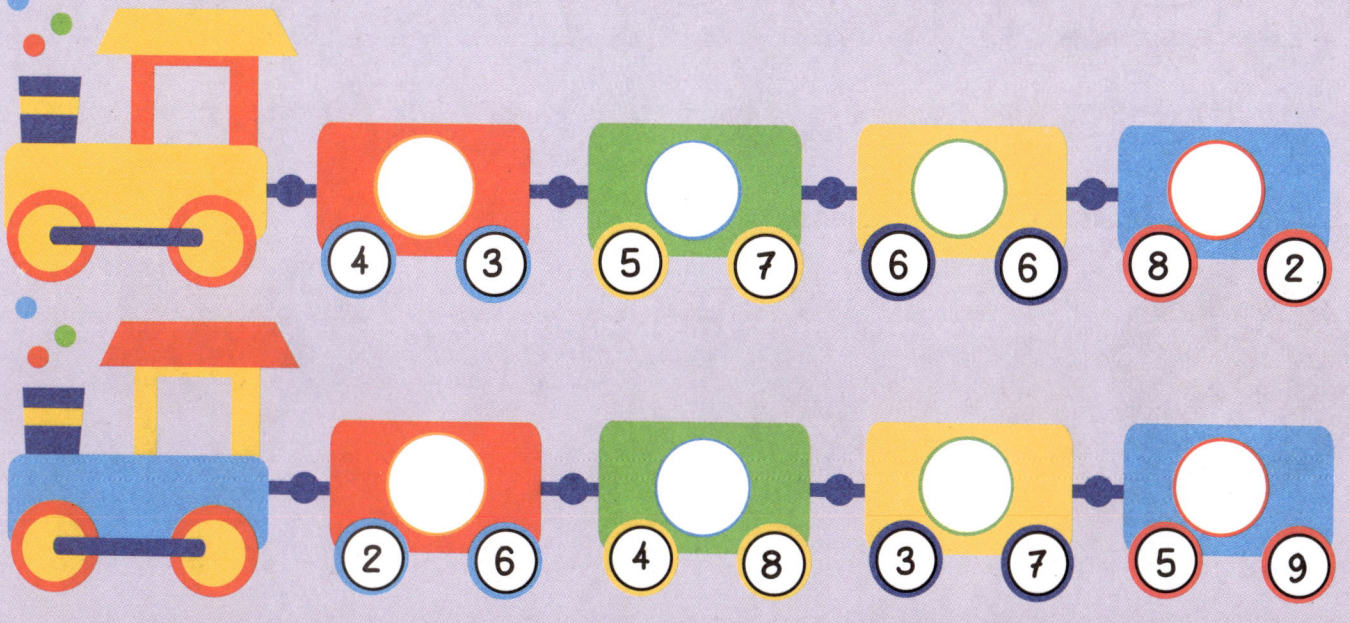

4 × 3   5 × 7   6 × 6   8 × 2

2 × 6   4 × 8   3 × 7   5 × 9

109. Solve and write the answers.

$16 \div 2 = \bigcirc$

$36 \div 4 = \bigcirc$

$27 \div 9 = \bigcirc$

$40 \div 5 = \bigcirc$

110. Color the numbers that are divisible by 2.

46	82	74	33	19
55	75	56	16	41
11	15	42	37	53
22	26	95	44	84

111. Solve and write the answers.

a. $81 \div 9 = \bigcirc$

b. $50 \div 2 = \bigcirc$

c. $6 \div 3 = \bigcirc$

112. Solve the divisions.

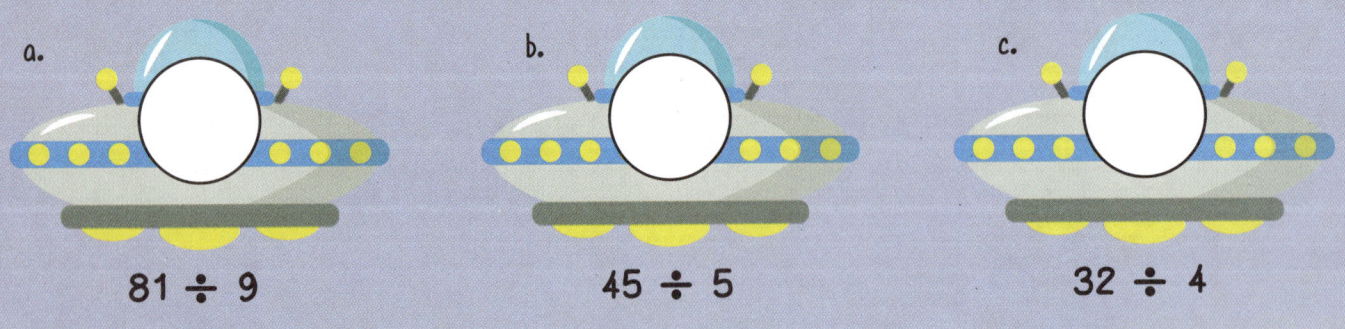

a. $81 \div 9$

b. $45 \div 5$

c. $32 \div 4$

113. Help the animals solve the divisions.

a. $42 \div 2$

b. $66 \div 3$

c. $42 \div 7$

114. Help the baby animals reach their food through maze by solving the divisions.

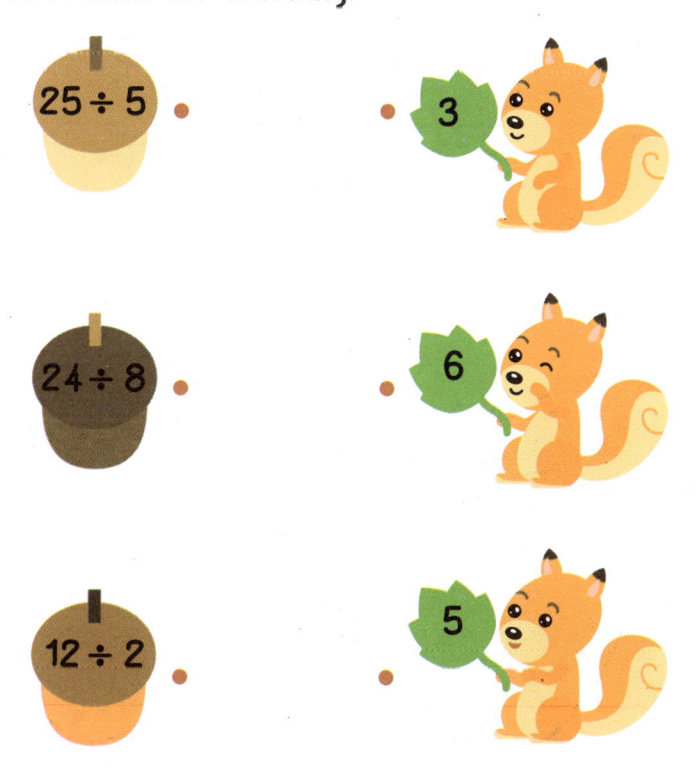

| 81 ÷ 9 | 24 ÷ 4 | 12 ÷ 4 | 49 ÷ 7 |

| 3 | 7 | 9 | 6 |

115. Match the following.

25 ÷ 5 •

24 ÷ 8 •

12 ÷ 2 •

• 3

• 6

• 5

116. Solve the division problems.

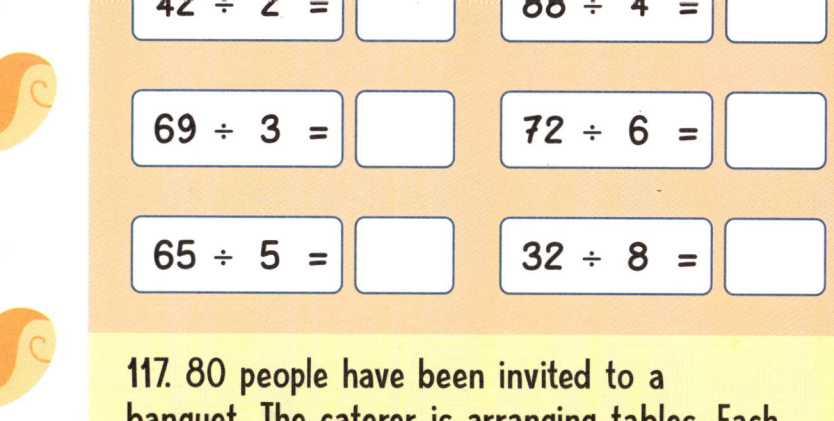

42 ÷ 2 =		88 ÷ 4 =	
69 ÷ 3 =		72 ÷ 6 =	
65 ÷ 5 =		32 ÷ 8 =	

117. 80 people have been invited to a banquet. The caterer is arranging tables. Each table can seat 10 people. How many tables are needed?

29

118. Count and multiply the given images.

$3 \times 3 =$ ☐

$4 \times 3 =$ ☐

$6 \times 2 =$ ☐

$5 \times 3 =$ ☐

119. Write the answers in the space provided.

_____ X _____ = _____

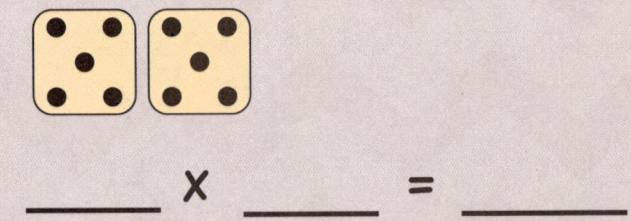

_____ X _____ = _____

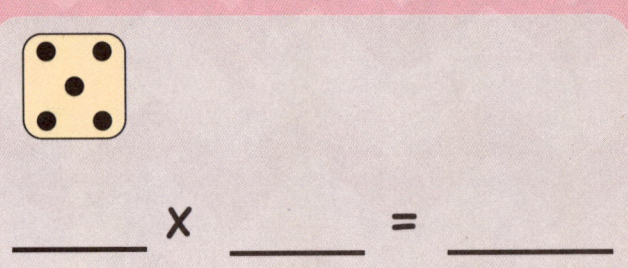

_____ X _____ = _____

120. Multiply each number with the number in the center to complete the web.

121. Color the apples that are multiples of 9.

4   16

9   36

122. Count and divide the given images. Write answers in the boxes.

123. Solve the following divisions.

124. Help the monkey eat the bananas by solving the divisions.

21 ÷ 3

14 ÷ 2

36 ÷ 4

125. Fill in the blanks with the correct numbers to get the given sum.

 (....) +  (10) + (15) =  (30)

 (9) +  (....) + (8) =  (30)

 (13) + (12) + (....) =  (32)

 (8) +  (11) + (....) =  (25)

 (....) + (3) + (8) = (19)

126. Fun with Numbers! Read the numbers and answer the following questions.

| 5 | 4 | 0 | 8 |

Make the smallest and largest four-digit number from the given numbers.

....

Make the smallest four-digit number that starts from 5.

....

127. Kacy bought a dress for $20 and a watch for $10. How much did she spend in all?

128. Miles had $60 in her pocket. She spent $50. How much money does she have now?

129. Ken got $40 from his mom, $20 from his dad and $10 from his brother. How much money he has in total?

130. Solve the following multiplications and have a fun day.

3 X 10

6 X 10

5 X 10

8 X 10

131. Identify the mathematical symbols and match them with their names.

 + •

 - •

 ÷ •

 X •

 > •

 < •

 % •

 = •

• Equal to

• Less than

• Addition

• Subtraction

• Percent

• Greater than

• Division

• Multiplication

132. Help Tina solve the following two-digit divisions.

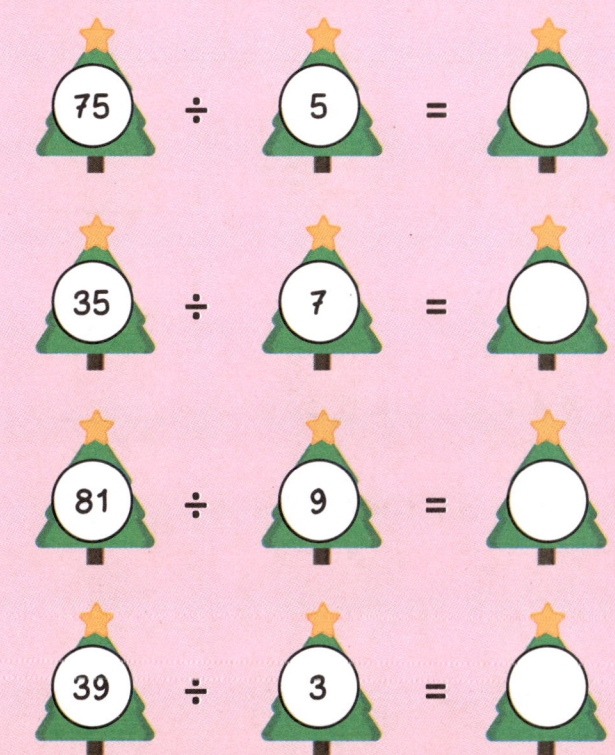

75 ÷ 5 =

35 ÷ 7 =

81 ÷ 9 =

39 ÷ 3 =

133. Select three pumpkins whose numbers add to give the sum equal to 30.

1   3

5   7

9   11

12   15

🍂 + 🍂 + 🍂 = 30

134. Write the numbers using the place values.

6 thousands + 5 hundred + 2 tens + 4 ones =

5 thousands + 0 hundreds + 1 tens + 7 ones =

9 thousands + 2 hundreds + 5 tens + 2 ones =

33

135. Fill in the boxes with the correct letters to get the names of the objects.

T=1   A=3   B=2   O=4

2	4	3	1

C=1   L=3   I=5 P=2   E=4   N=6

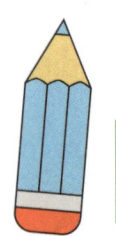

2	4	6	1	5	3

136. Write eight hundred and fifty-four in numbers.

_____

137. Subtract 9 from 20

_____

138. Complete the pattern:
3, 6, 9 __ , __ 18, 21, __ , __ , 30

_____

139. What comes after 101?

_____

140. Read carefully and fill in the blanks.

• 4 x 5 = 20, then 20 ÷ 4 = _____

• 7 x 3 = 21, then 21 ÷ 7 = _____

• 9 x 4 = 36, then 36 ÷ 9 = _____

• 4 x 8 = 32, then 32 ÷ 4 = _____

• 1 x 6 = 6, then 6 ÷ 1 = _____

• 5 x 5 = 25, then 25 ÷ 5 = _____

141. Help Matt solve the following two-digit divisions.

 ÷  =

 ÷  =

 ÷  =

142. It's raining numbers. Arrange them in ascending order in the box given below.

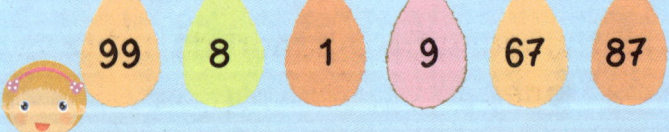

143. Color the dices to get the sum equal to the given numbers.

    7

    12

    9

    4

34

144. Answer the questions by choosing the correct number.

1000	639	700	220	485

Number between 950 and 1050.

Number between 600 and 650.

Number you get if you multiply 7 by 100.

Number smaller than 300.

Number bigger than 400 but smaller than 490.

145. Use >, < or = to compare the numbers.

62		39	29		29
72		44	38		76
43		90	12		21
22		10	18		19
9		19	5		5
88		99	89		98

146. Solve the circular addition puzzle.

5	+	....	=	12
+		+		+
8	+	3	=	....
=		=		=
....	+	10	=	23

147. Arrange the hats in descending order.

909 450 1001 567 234 893 111

----

148. Arrange the hats in ascending order.

967 666 546 985 232 1010 1020

----

149. Color the fish with the greater number.

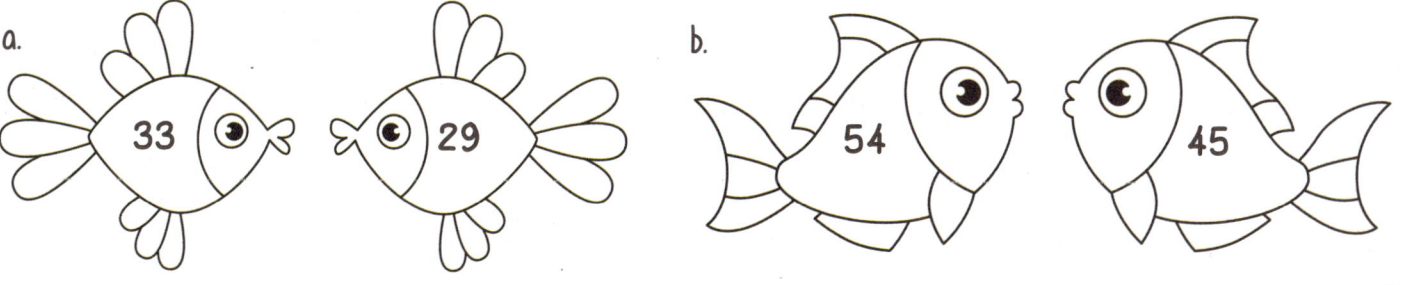

a.   33   29

b.   54   45

150. What's the time on the clocks?

a.

b.

c.

d.

151. Draw hands on the clocks to show the time. Write the time in figures in the boxes.

a. Six o'clock

`06:00`

b. Seven o'clock

c. Twelve o'clock

d. Three o'clock

152. Show time on the clocks that the snails are carrying on their backs.

2:00

5:30

8:35

153. Help Dylan get his treats by drawing hands on the clocks.

a.  Four o'clock

b. Quarter past nine

c. Half past seven

d. Quarter to one

154. Show your mealtimes on the clocks. Write time in the blank space.

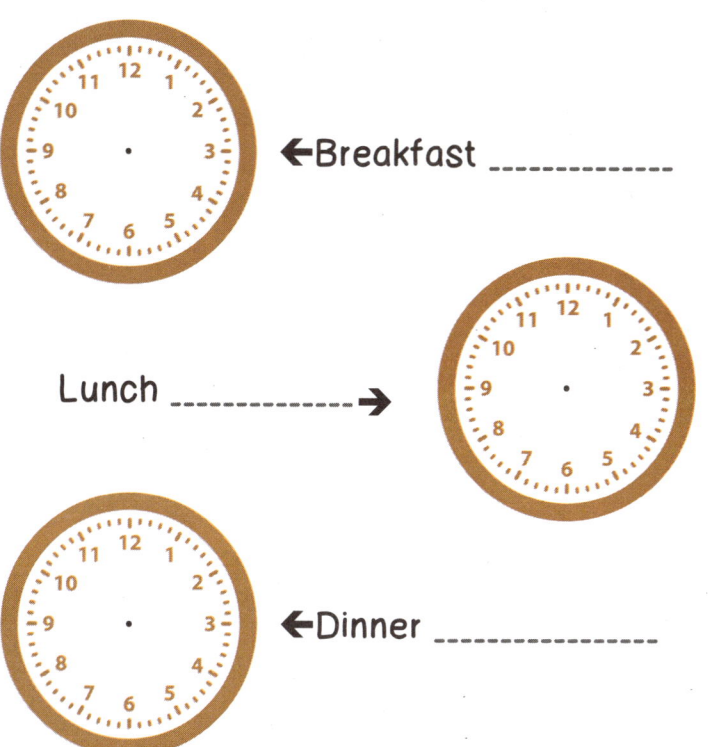

←Breakfast _____

Lunch _____→

←Dinner _____

155. Write the answers in numbers.

	seconds in one minute
	minutes in one hour
	hours in one day
	days in one week
	weeks in one year
	months in one year
	days in one year
	seasons in one year

156. Match the numbers on the clock.

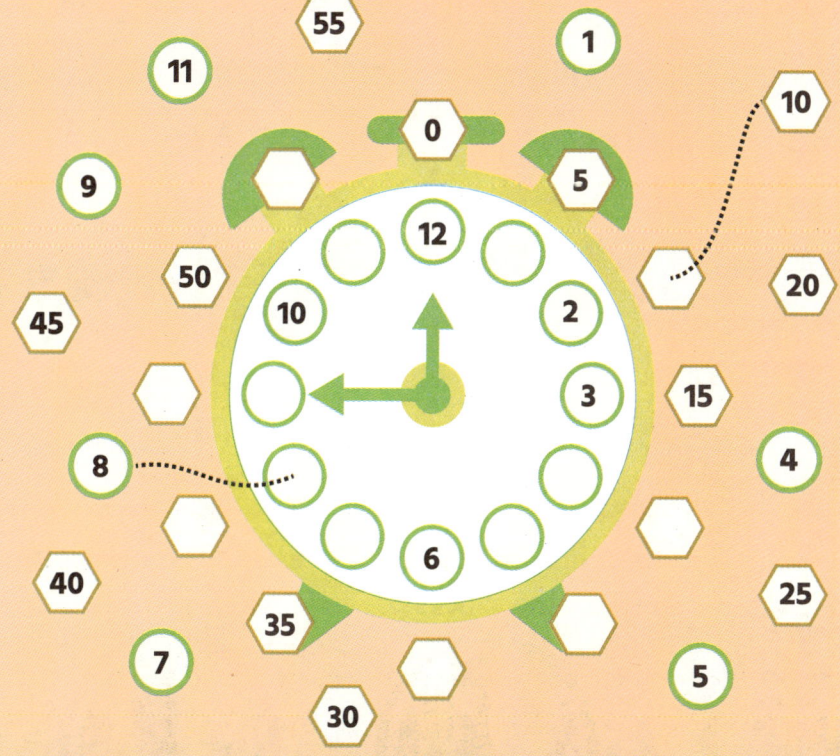

157. Tick the correct time.

a.

○ 6 o'clock
○ 8 o'clock
○ 4 o'clock

b.

○ 1 o'clock
○ 3 o'clock
○ 5 o'clock

158. My swimming class starts at 4 pm and ends at 6 pm. What is the duration of the class?

159. Read both the hands in the clock and write the time in both ways.

Quarter to

_____

_____

160. Answer the following questions.

a. The clock is 30 minutes forward. What is the right time?

b. The clock is 45 minutes behind. What is the correct time?

----

161. How much time has passed? Think and write.

6:15 pm to 9:15 pm — 3 hours

8:30 am to 12:00 pm —

12:20 pm. to 1:00 pm —

9:50 am to 11:40 am —

5:00 pm to 5:45 pm —

3:30 pm to 6:30 pm —

162. Match the time in numbers with words.

 4:15 • • Five past two.

 3:30 • • Quarter past four.

 2:05 • • Nine o'clock.

 7:45 • • Five to five.

 4:55 • • Half past three.

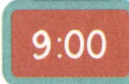 9:00 • • Quarter to eight.

163. Fill in the blanks.

_____ seconds = 1 minute

60 minutes = __ hour

__ days = 1 week

12 _____ = 1 year

____ days = 1 year

4 weeks = __ month

164. How many minutes are shown in colored part of the clocks?

a.

b.

165. What's the time on the clocks?

a.  Quarter past _____

_____

:

b. _____

_____

:

c. _____

_____

:

166. Match the digital clocks and wall clocks.

 **1 : 00** •

 •

10 : 15 •

•

7 : 25 •

•

5 : 30 •

•

167. Draw the minute and hour hands and write the timings of your daily routine.

a. I wake up at

b. I take a bath at

c. I eat breakfast at

d. I go to school at

e. I eat lunch at

f. I do my homework at

g. I play at

h. I eat dinner at

i. I sleep at

168. Choose the correct fraction.

a. 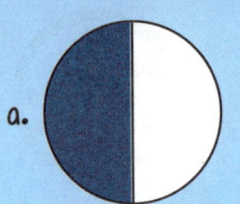 $\frac{5}{8}$ $\frac{1}{2}$ $\frac{4}{3}$

d. 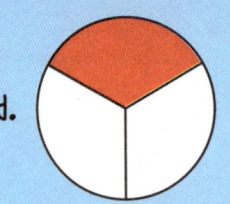 $\frac{9}{2}$ $\frac{1}{3}$ $\frac{6}{8}$

b.  $\frac{4}{12}$ $\frac{13}{5}$ $\frac{3}{8}$

e. 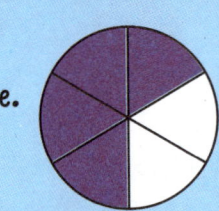 $\frac{5}{8}$ $\frac{5}{6}$ $\frac{4}{6}$

c. 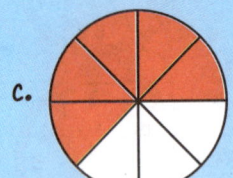 $\frac{4}{8}$ $\frac{5}{8}$ $\frac{6}{8}$

f. 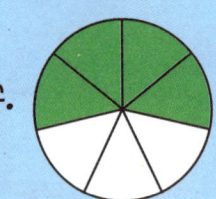 $\frac{4}{7}$ $\frac{5}{7}$ $\frac{1}{7}$

169. Color and write the fraction.

a.

b.

c.

170. Write the fraction of the colored fruit seeds.

a.

b.

171. Write fractions of the white part of the fruit.

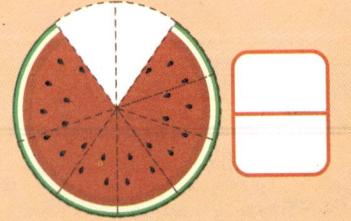 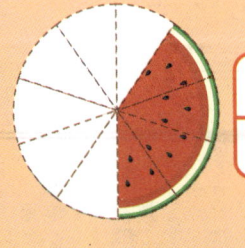 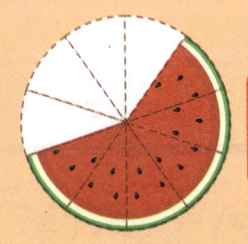 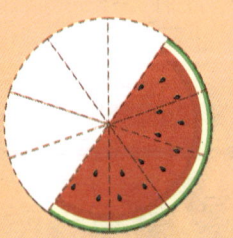

172. Solve the fraction sums.

a.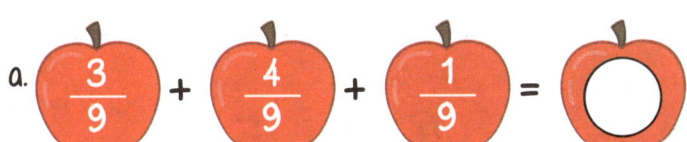

b.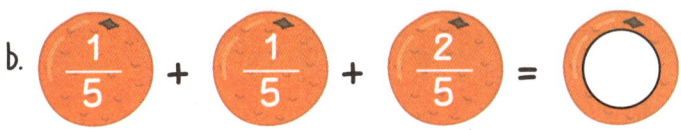

c.

173. There are 10 skirts on the rack. Gilly took 3. What fraction of the skirts did she take?

 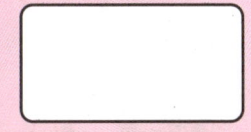

174. Jane saw 7 starfish and brought 2 home with her. What fraction of the starfish did she bring home?

175. Color the fruits according to the given fractions.

a. $\dfrac{5}{7}$

b. $\dfrac{2}{5}$

176. Solve the fractions and color the rings accordingly.

 $\dfrac{1}{6} + \dfrac{2}{6}$

 $\dfrac{4}{6} - \dfrac{1}{6}$

 $\dfrac{5}{6} - \dfrac{1}{6}$

177. Compare the fractions and mark greater than or less than (<, >).

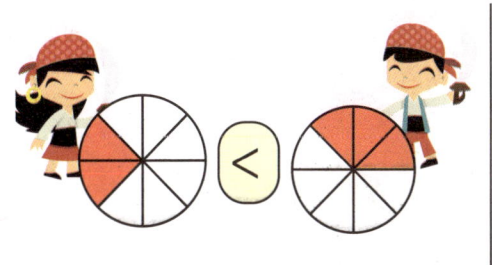 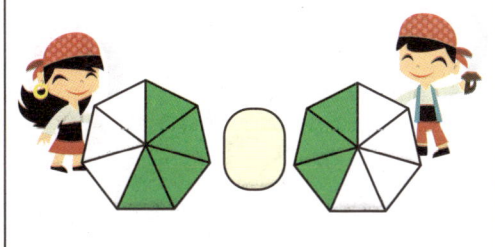 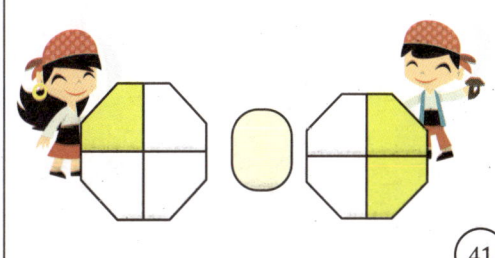

178. See the price on the tag. Count the coins and mark if you can buy the item.

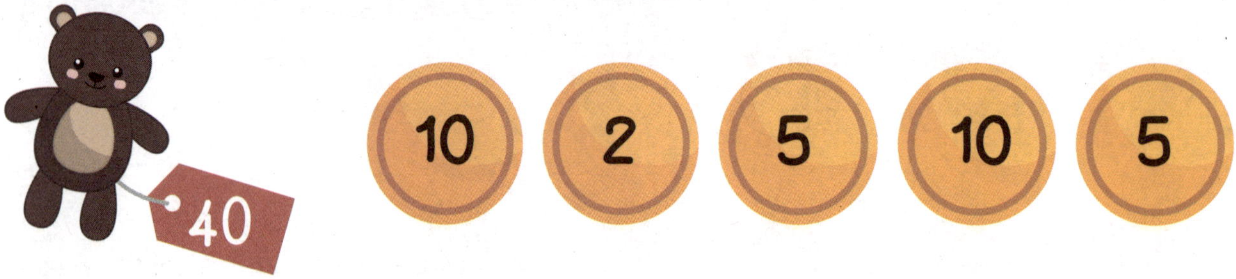

🧸 **40**	10	2	5	10	5	○ Yes ○ No
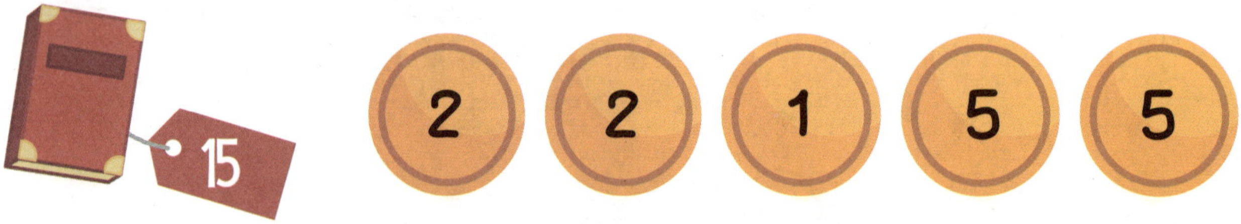 📕 **15**	2	2	1	5	5	○ Yes ○ No
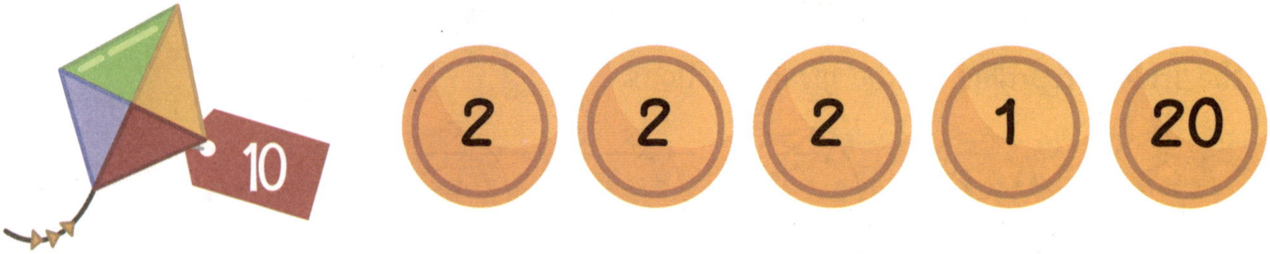 🪁 **10**	2	2	2	1	20	○ Yes ○ No
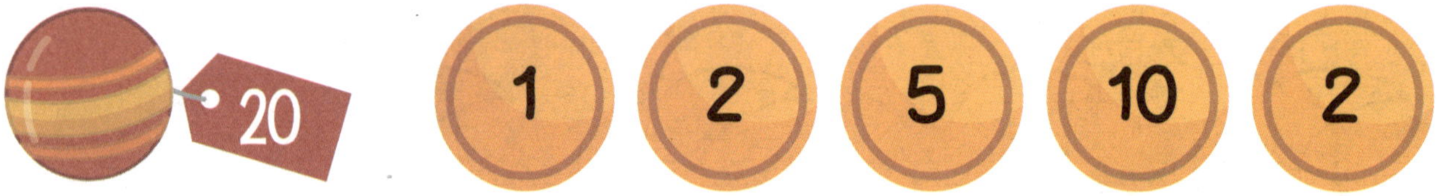 ⭕ **20**	1	2	5	10	2	○ Yes ○ No

179. Count the value of coins and write the answer.

180. Add the money and write the price on tags.

a.

b.

c.

181. Write the total by adding the value given to each object.

8   5   3   4   6   2   7   3

What is the total?

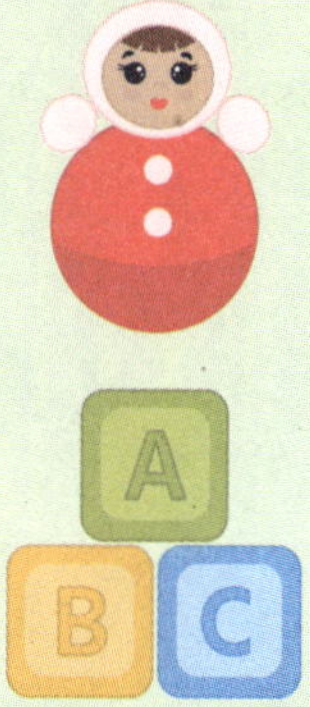

```
- - - - - - - -
+
- - - - - - - -

- - - - - - - -
```

What is the total?

```
- - - - - - - -
+
- - - - - - - -

- - - - - - - -
```

What is the total?

```
- - - - - - - -
+
- - - - - - - -
```

What is the total?

```
- - - - - - - -
+
- - - - - - - -
```

---

182. Mike bought a watch for $200 and a ring for $100. How much did he spend in all?

183. Haley had $600 in her pocket. She spent $50 on a pair of socks. How much money does she have now?

184. Ken saves $20 every month. How much money does he have in total at the end of the year?

185. Measure the height and write it in the box.

a. [ ] centimeters

b. [ ] centimeters

c. [ ] centimeters

186. If 1 feet=12 inches, convert and write.

a. **3** ft = ( ) in   b. **5** ft = ( ) in   c. **2** ft = ( ) in   d. **6** ft = ( ) in

187. 1 kg=1000 g. Solve the following accordingly.

5 kg + 2 kg = [ ] g

9 kg + 400 g = [ ] g

2500 g + 100 g = [ ] g

1.5 kg + 200 g = [ ] g

3 kg + 400 g = [ ] g

7 kg + 50 g = [ ] g

188. Write the capacity.

160ml = [four cups]   [one cup] [ ]

300ml = [six cups]   [one cup] [ ]

44

189. Fill in the numbers such that the sum of each side is equal to the other.

190. Calculate the total weight by using the value given for each fruit.

🍎 20g   🍐 30g   🥝 10g

a.   ___ g

b.   ___ g

c.   ___ g

d.   ___ g

e.   ___ g

f.   ___ g

45

191. Color the glasses that can be filled with the juice in the jug.

= 100ml

400ml

500ml

192. Arrange the given numbers in ascending order.

| 327 | 382 |
| 361 | 312 |

| 618 | 655 |
| 624 | 611 |

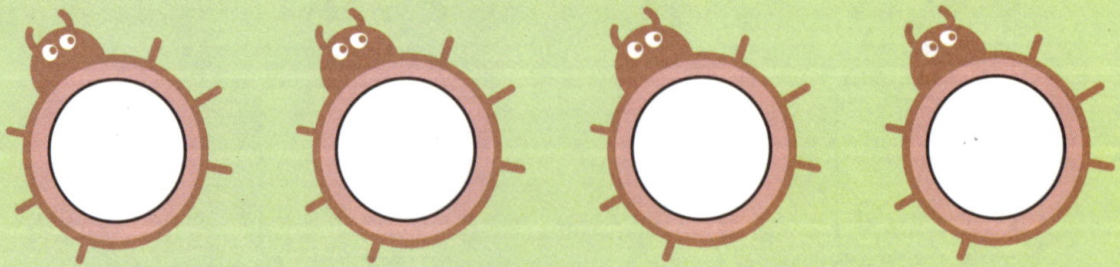

193. Solve and write answers in the boxes.

6x5

9-2

4+4

8÷2

46

194. Answer the following questions with reference to the clock.

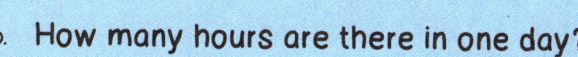

a. What is the time on the clock?

_ _ _ _ _ _ _ _ _ _ _ _ _ _ _ _ _ _

b. How many hours are there in one day?

_ _ _ _ _ _ _ _ _ _ _ _ _ _ _ _ _ _

c. What is the half of 12 hours?

_ _ _ _ _ _ _ _ _ _ _ _ _ _ _ _ _ _

195. If the burger is 150g and the apple is 15g.
How many apples will be equal to the weight of 1 burger?

196. Look at the teapot and write the number on the label.

6 tens + 4 ones =

4 tens + 8 ones =

5 tens + 1 ones =

3 tens + 8 ones =

3 tens + 2 ones =

197. Use the value given to each monster to solve the sums.

3   2   5

⬡ + ⬡ + ⬡ = ☐

⬡ + ⬡ + ⬡ = ☐

⬡ + ⬡ + ⬡ = ☐

⬡ + ⬡ + ⬡ = ☐

198. Color according to the given fractions.

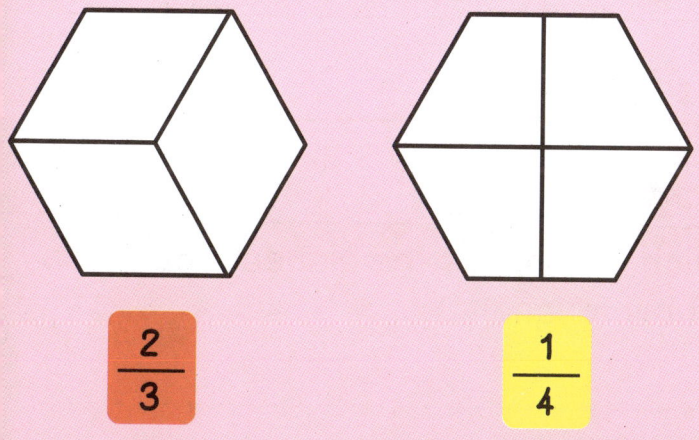

$\dfrac{2}{3}$

$\dfrac{1}{4}$

199. Subtract the numbers on top and write the answers in the bubble below.

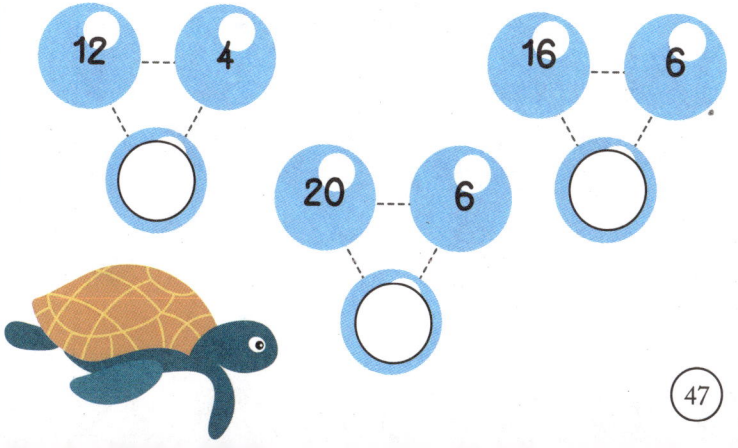

12 --- 4   16 --- 6

○   20 --- 6   ○

○

47

200. Study the graph and answer the questions that follow.

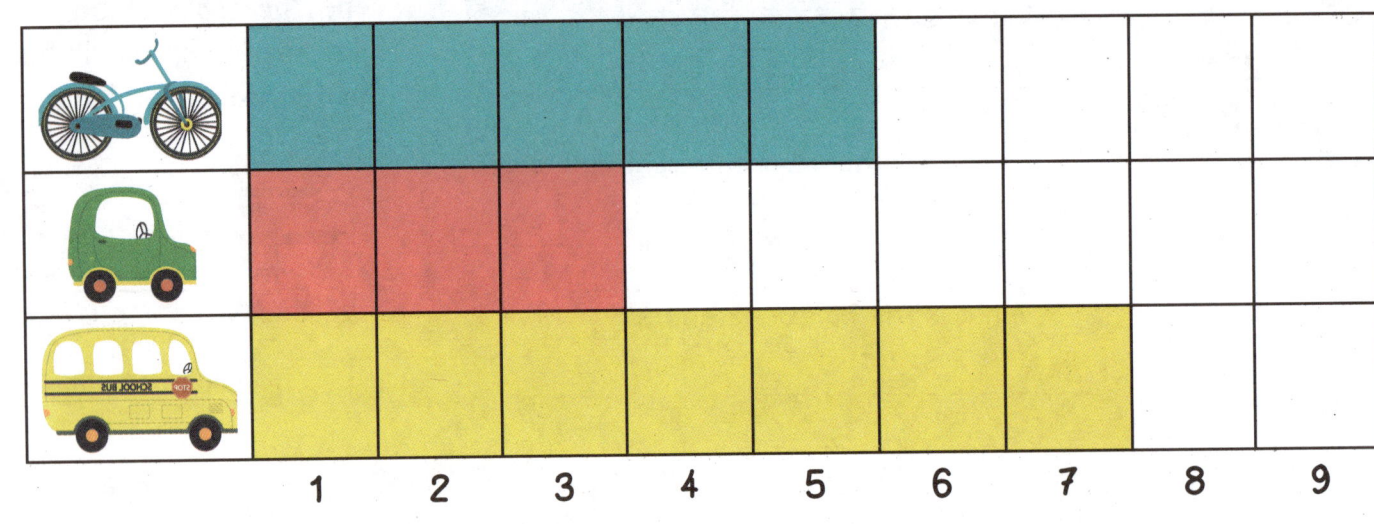

Transportation

1   2   3   4   5   6   7   8   9

Number of children

a. How many kids take the bus to school?

b. How many kids ride a bike to school?

c. How many kids go by car to school?

d. How many kids ride in the car and bus in total?

201. Tally the sea animals and answer the question accordingly.

a. 	
b.	
c.	
d.	

e. Which animal has the highest count and by how much?

_ _